FLORA OF TROPICAL EAST AFRICA

GERANIACEAE

J. O. KOKWARO
(University College Nairobi)*

Herbs, shrubs, or suffrutices, very rarely arborescent; stems fleshy. Leaves alternate or opposite, radical ones occasionally in basal rosettes, if opposite often unequal, usually stipulate and petiolate, serrate, crenate or dentate, lobed, dissected or compound, rarely entire. Inflorescence usually axillary, sometimes pseudumbellate, occasionally 1-flowered, determinate, bracteate. Flowers ☿ or very rarely dioecious, usually chasmogamous but occasionally cleistogamous, actinomorphic or zygomorphic, hypogynous, generally (4)5(8)-merous, heterochlamydous. Sepals (4)5, free or connate at the base, imbricate or rarely valvate, the posterior (uppermost) one sometimes spurred (especially in *Pelargonium*). Petals 5 (rarely 8, 4, 2, or even 0 by reduction), free, mostly unequal, hypogynous or subperigynous, imbricate or rarely contorted. Disk (extrastaminal or nectariferous) glands often present, alternating with the petals. Stamens twice as many as the sepals, more rarely three times as many, some occasionally sterile; filaments free, usually ± connate at the base, sometimes connate in 5 bundles of 3 each; anthers versatile, introrse, 2-thecous, dehiscing longitudinally; pollen grains of different types, usually 3-colpate or 3-colporate. Ovary superior, syncarpous, (2)3–5(8)-locular; ovules 1–2 in each chamber, pendulous, anatropous, superposed; placentation axile; style present and adhering to the beak or absent; stigmas ligulate, clavate or filiform, rarely capitate. Fruit a schizocarp or sometimes a 3–5(rarely 8)-lobed capsule; mericarps rostrate, breaking away from a persisting central column, usually 1-seeded (rarely 2–many-seeded) and dehiscing septicidally. Seeds smooth or minutely reticulate; embryo curved, rarely straight; endosperm scanty or absent.

A family of 5 genera, namely *Erodium* Ait., *Geranium*, *Monsonia*, *Pelargonium*, and *Sarcocaulon* (DC.) Sweet, all occurring in Africa, and about 800 species widely distributed throughout temperate and subtropical regions of both hemispheres. The economic value of the family is based primarily on the florist's " geraniums ", as which *Pelargonium* cultivars are widely known; some essential oils can also be obtained from species of *Pelargonium*, notably *P. graveolens* Ait.

Flowers actinomorphic; sepals not spurred; extrastaminal
    glands present:
    Stamens 10; schizocarp 1–3·5 cm. long; rostrum of
      mericarps rolling upwards at dehiscence   . . **1. Geranium**
    Stamens 15; schizocarp 5–11 cm. long; rostrum of
      mericarps twisting helically at dehiscence . . **2. Monsonia**
Flowers zygomorphic; posterior sepal with a spur adnate to
    the pedicel; extrastaminal glands absent; stamens 10,
    but only 2–7 bearing anthers .   .   .   . **3. Pelargonium**

1. GERANIUM

L., Sp. Pl.: 676 (1753) & Gen. Pl., ed. 5: 306 (1754)

Annual or perennial herbs, rarely subshrubs, sometimes tufted, occasionally woody at the base. Leaves alternate or opposite, petiolate, palmately

* Account prepared as part of a course of studies at the University of Uppsala, Sweden.

lobed or dissected, rarely toothed. Inflorescence a simple axillary umbel of 2–3 flowers or the flowers solitary, the peduncle occasionally very short or lacking. Flowers actinomorphic. Sepals 5, imbricate, usually awn-tipped. Petals 5, mostly notched, hypogynous, imbricate. Stamens 10, usually all fertile, very rarely 5 without anthers; filaments free or shortly connate at the base. Extrastaminal glands 5 and alternating with the petals, generally located at the base of the stamens. Ovary 5-lobed and 5-locular, usually long beaked, each locule containing 2 superposed ovules, 1 ovule aborting; style glabrous inside, adhering to the beak, recoiled at maturity but not twisting; stigmas 5, filiform. Fruit a schizocarp containing 4–5 mericarps, dehiscing by the splitting of the rostrum, the segments of which roll upwards, remaining attached at the apex, thus releasing the 1-seeded mericarp from the base of the calyx; in many species the mericarps open at the same time and the seeds are shot out. Seeds with thin endosperm or exalbuminous; embryo with leafy cotyledon, curved or folded.

About 400 species, cosmopolitan but mostly in the temperate regions and to a lesser extent in the subtropical regions of both hemispheres.

Stems with sharp prickles; leaf-lamina divided to
    about three-quarters of the radius   .   .   1. *G. aculeolatum*
Stems without prickles:
  Stipules bifid nearly to the base:
    Leaf-lamina divided to more than three-
      quarters of the radius; corolla without a
      dark centre; mericarps smooth:
      Leaves softly white tomentose beneath   .   2. *G. incanum*
      Leaves coarsely pilose beneath, especially
        along the veins   .   .   .   .   3. *G. vagans*
    Leaf-lamina divided to less than three-
      quarters of the radius; corolla with a
      darker purple centre; mericarps reticu-
      lately ridged   .   .   .   .   4. *G. ocellatum*
  Stipules entire:
    Plant annual; stipules ovate-lanceolate, 0·3–
      4 mm. long, less than 2 mm. wide; meri-
      carps with shallow reticulate ridges   .   5. *G. elamellatum*
    Plant perennial; stipules ovate, 5–12 mm.
      long, 3–5 mm. wide; mericarps smooth:
      Leaf-lamina reniform; flowers usually
        solitary .   .   .   .   .   6. *G. kilimandscharicum*
      Leaf-lamina pentagonal; flowers usually in
      pairs   .   .   .   .   .   7. *G. arabicum*

1. **G. aculeolatum** *Oliv.*, F.T.A. 1 : 291 (1868); Engl., Hochgebirgsfl. Trop. Afr. : 274 (1892); R. Knuth in E.P. IV. 129 : 203, fig. 26/C–F (1912); F.P.S. 1 : 128 (1950); Milne-Redh. in Mem. N.Y. Bot. Gard. 8 : 230 (1953); E.P.A. : 345 (1956); Petit in F.C.B. 7 : 23 (1958); F.P.U. : 63 (1962); Laundon in F.Z. 2 : 131, t. 20/E (1963). Type: Ethiopia, Mt. Jumetiz, Maitza Kola, *Schimper* 665 (K, lecto. !, P, isolecto. !)

Perennial herb up to 9 dm. long, climbing or trailing, rooting at the nodes; stems herbaceous, longitudinally grooved, armed with sharp reflexed prickles, and with glandular hairs on the younger parts; internodes 2–15 cm. long, 1–3 mm. thick, sometimes pinkish especially the older parts. Leaves opposite and palmately lobed; lamina 5-lobed, divided to about three-quarters of the radius; lobes 2–12 mm. wide at the base, rhombic in outline, pinnatifid, stiffly pilose on both surfaces, particularly on the veins of the lower surface;

petiole 1–10 cm. long, longitudinally grooved, covered with sharp reflexed prickles and often with glandular hairs; stipules divided nearly to the base into lanceolate to linear segments, 3–8 mm. long, often glandular hairy towards the margin, margin often hyaline. Inflorescence 2-flowered; peduncle 2–10 cm. long, grooved or terete, sometimes with prickles, usually densely glandular; pedicels 5–25 mm. long; bracts linear-lanceolate, 2–5 mm. long, hyaline along the edges, mostly stiffly pilose. Sepals lanceolate, 7–8 mm. long, 1–4 mm. wide, mucronate, 3-nerved, hyaline at the edge, usually stiffly pilose outside but glabrous inside. Petals broadly spathulate, 7–10 mm. long, ± 5 mm. wide, obtuse, white or pale pink to pale mauve. Stamens: filaments narrowly lanceolate, 4–5 mm. long, scarcely hispid towards the base along the edges; anthers 0·8–1·1 mm. long, 0·5–0·7 mm. wide. Ovary pubescent; stigmas 1·5–2·5 mm. long, slightly hispid or pubescent. Mericarps smooth, pilose; mature rostrum 15–25 mm. long, 1–1·5 mm. wide, shortly pubescent. Seeds 2–2·8 mm. long, 1–1·4 mm. wide, glabrous; testa scarcely reticulate.

UGANDA. Ruwenzori, Ibonde Pass, 1 Oct. 1932, *A. S. Thomas* 769!; Kigezi District: Kachwekano Farm, July 1949, *Purseglove* 3030! & Mafuga Forest, 16 May 1951, *Dawkins* 732!
KENYA. Northern Frontier Province: Mt. Nyiru, 25 July 1960, *Kerfoot* 2038!; Naivasha District: Kinangop, 21 Dec. 1930, *Napier* 702!; Embu District: Gathiba R., Sept. 1956, *Ossent* 193!
TANGANYIKA. Arusha District: Mt. Meru, W. side, 21 July 1957, *Bally* 11578!; Ufipa District: Luiche valley near Sumbawanga, 17 Apr. 1934, *Michelmore* 1033!; Songea District: Matengo Hills, Kihuru Hill near Ngwambo, 22 May 1956, *Milne-Redhead & Taylor* 10414!
DISTR. **U**2, 3; **K**1–5; **T**2–4, 6–8; Congo and Sudan Republics, Ethiopia, Malawi and Mozambique
HAB. Upland rain-forest, common at edges, in clearings, along streams and in marshy places, sometimes persisting as a weed in cultivation; 1200–3050 m.

2. **G. incanum** *Burm. f.*, Spec. Bot. Geraniis: 28, t. 1/26 (1759); L., Sp. Pl., ed. 2: 957 (1763); R. Knuth in E.P. IV. 129: 162, fig. 23/F–H (1912); Salter in Fl. Cap. Pen.: 509 (1950); Laundon in Bol. Soc. Brot., sér. 2, 36: 63 (1961) & in F.Z. 2: 134 (1963). Type: South Africa, Cape of Good Hope, *Oldenland* (Herb. Oldenland, G, holo.)

Perennial semiscandent, creeping and scrambling herb up to 1 m. long; stems longitudinally grooved, pilose (or glabrous below) but sometimes becoming nearly tomentose or densely glandular on the upper parts. Leaves: lowermost in basal rosettes, middle ones alternate or opposite, and the terminal sets usually opposite; lamina 3–7-lobed nearly to the base; lobes 0·5–2 mm. wide at the base, ovate to elliptic in outline, deeply divided into linear or oblong-lanceolate segments, pilose above and silvery tomentose beneath; petiole 0·5–25 cm. long, longitudinally grooved, pilose; stipules divided nearly to the base into linear-lanceolate segments, 3–15 mm. long, pilose especially on the abaxial surface. Inflorescence 2-flowered; peduncle 1–3 cm. long, pilose to tomentose or glandular; pedicels 1–4 cm. long; bracts linear-lanceolate, 2–7 mm. long, hyaline along the edges, pilose or tomentose. Sepals ovate-lanceolate, 5–10 mm. long, 2–3 mm. wide, cuspidate, 3-nerved and pilose, hispid to tomentose and with glandular hairs outside. Petals obovate, 6–15 mm. long, 4–8 mm. wide, retuse, white, pink or mauve with dark veins. Stamens: filaments narrowly lanceolate, 4·2–6 mm. long, hispid to pubescent at base, cream or brown; anthers 1·2–1·5 mm. long, 0·6–0·8 mm. wide, yellowish-brown. Ovary pubescent; stigmas 1–4 mm. long, red or pale pink. Mericarps smooth, pilose; mature rostrum 13–27 mm. long, 1–2 mm. wide, pubescent or glandular. Seeds 1·6–2 mm. long, 0·9–1·2 mm. wide, glabrous; testa covered with a shallow reticulate pattern, pale pink to dark brown.

subsp. **incanum**

Leaf-segments linear; glandular hairs on the younger parts of the stems containing dark red pigments.

Tanganyika. Iringa District: Dabaga Highlands, Kibengu, 16 Feb. 1962, *Polhill & Paulo* 1500!; Njombe District: Mdapo, Mar. 1954, *Semsei* 1658! & Elton Plateau, 7 Jan. 1957, *Richards* 7563!
Distr. **T7**; South Africa
Hab. Upland grassland; 1500–2800 m.

Syn. *G. multisectum* N.E. Br. in K.B. 1901: 120 (1901). Type: Lesotho, Machacha Mt., *Bryce* (K, holo.!)

subsp. **nyassense** (*Knuth*) *Laundon* in Bol. Soc. Brot., sér. 2, 35: 63, t. 3 (1961) & in F.Z. 2: 134, t. 20/B (1963). Type: Tanganyika, Mt. Rungwe, *Stolz* 1389 (B, holo. †, K, iso.!)

Leaf-segments oblong-lanceolate; glandular hairs on younger parts of the stems containing brownish pigments.

Tanganyika. Iringa District: Mufindi West, 31 Jan. 1934, *Michelmore* 940!; Rungwe District: Ukinga Mts., Mwakaleli [Mwakalila], 31 Mar. 1914, *Stolz* 2599! & W. Porotos [Mporotos] to Mt. Rungwe, 13 Aug. 1933, *Greenway* 3550!
Distr. **T7, 8**; Malawi, Mozambique and Rhodesia
Hab. Upland grassland, often along streams or in marshy places; 900–2600 m.

Syn. *G. nyassense* Knuth in F.R. 18: 289 (1922)
      G. ukingense Knuth in F.R. 18: 292 (1922). Type: Tanganyika, Ukinga Mts., *Stolz* 2060 (B, holo. †, BM, iso.!)

Variation (of species as a whole). There is considerable variation within *G. incanum* especially in the leaf-shape, indumentum, size of the petals, rostrum and stigmas. See discussion by Laundon in Bol. Soc. Brot., sér. 2, 35: 63–67 (1961).

3. **G. vagans** *Bak.* in K.B. 1897: 246 (1897); R. Knuth in E.P. IV. 129: 208 (1912); Milne-Redh. in Mem. N.Y. Bot. Gard. 8: 231 (1953); A.V.P.: 123 & 293 (1957); Petit in F.C.B. 7: 23 (1958); Laundon in Bol. Soc. Brot., sér. 2, 35: 71 (1961) & in F.Z. 2: 134, t. 20/C (1963). Type: Malawi, Nyika Plateau, *Whyte* (K, holo.!)

Perennial semi-procumbent, creeping and climbing herb 3–10 dm. long; stems herbaceous, longitudinally grooved, pilose on the older parts but usually densely glandular on the younger parts; internodes 1–14 cm. long, 1·5–3 mm. thick, frequently pinkish-red or dull purple. Leaves alternate on the lower portion of the stem, opposite towards the apex, sometimes only in basal rosettes, star-like in outline; lamina 5-lobed, divided up to about nine-tenths of the radius; lobes 1–2 mm. wide at the base, ovate in outline, deeply divided into linear or oblong-lanceolate segments, stiffly pilose above as well as on the veins beneath and along the margins, otherwise glabrous on the undersurface; petiole 0·5–21 cm. long, terete or longitudinally grooved, pilose or glandular; stipules divided nearly to the base into lanceolate segments, 3–15 mm. long, pilose or glandular, usually pilose and reddish on the abaxial surface. Inflorescence 2-flowered; peduncle (1–)3–25 cm. long, pilose or glandular, rarely glabrous; pedicels 1–8 cm. long; bracts narrowly lanceolate, 3–15 mm. long, pubescent. Sepals elliptic-lanceolate, 6–8 mm. long, 2–3 mm. wide, cuspidate, 3-nerved, pubescent, pilose or glandular on the outer surface, glabrous on the inner surface, red and greenish in colour. Petals broadly spathulate, 10–12 mm. long, 5–7 mm. wide, obtuse, pink or mauve, rarely white, clearly nerved. Stamens: filaments narrowly lanceolate, 4·5–6 mm. long, cream, hispid along the edges towards the base; anthers 1·1–1·7 mm. long, 0·6–0·8 mm. wide, brown. Ovary glandular or pilose; stigmas 2–4 mm. long, purple or red. Mericarps smooth, pilose; mature rostrum 15–20 mm. long, 1·0–1·5 mm. wide, glandular or pubescent. Seeds

2·3–2·7 mm. long, 1·7–2 mm. wide, glabrous; testa covered with a shallowly reticulate pattern, dark brown.

subsp. **vagans**

Rostrum of fruit covered with long glands.

UGANDA. Kigezi District: Muhavura Mt., 7 Dec. 1930, *B. D. Burtt* 2820!; Elgon, Mudangi [Madangi], 5 Sept. 1932, *A. S. Thomas* 596!
KENYA. Elgon, eastern slope above Tweedie's saw-mill, 9 May 1948, *Hedberg* 844!; Naivasha District: Kinangop, 21 Dec. 1930, *Napier* 726!; Mt. Kenya, Soames' camp, 28 Apr. 1942, *MacLaughlin* in *Bally* 2801!
TANGANYIKA. Kilimanjaro, 1 Jan. 1913, *Grote* 3943!; Morogoro District: Uluguru Mts., Lukwangule Plateau, 3 Jan. 1934, *Michelmore* 889!; Rungwe District: foothills of Mt. Rungwe, Kiwira, May 1953, *Eggeling* 6516!
DISTR. **U**2, 3; **K**3, 4; **T**2, 6, 7; Congo Republic and Malawi
HAB. Upland grassland and moor, also extending into upland rain-forest, often along streams or in marshy places; 1370–4500 m.

SYN. *G. angustisectum* Knuth in E.P. IV. 129: 207 (1912). Type: Tanganyika, Kilimanjaro, above Kibosho Forest, *Volkens* 1534 (B, holo. †, BM, iso.!)
     G. schliebenii Knuth in N.B.G.B. 11: 1067 (1934). Type: Tanganyika, Uluguru Mts., *Schlieben* 3486 (B, holo. †, BM, BR, P, S, iso.!)

NOTE. Subsp. *whytei* (Bak.) Laundon, which has the rostrum of the fruit densely covered in hairs but without glands, is known only from the mountains of southern Malawi.

4. **G. ocellatum** *Cambess.* in Jacquem., Voy. Ind. 4, Bot.: 33 (1844) & Atlas 2, t. 38 (1844); Engl., Hochgebirgsfl. Trop. Afr.: 274 (1892); R. Knuth in E.P. IV. 129: 62 (1912); Milne-Redh. in F.W.T.A., ed. 2, 1: 157 (1954); E.P.A.: 346 (1956); Laundon in F.Z. 2: 136 (1963). Type: Pakistan (or Kashmir), " Pentapotamia ", near Sera, *Jacquemont* 203 (P, lecto.!)

Annual prostrate-ascending herb up to 7 dm. long/tall, diffusely branched from below; stems longitudinally grooved, mostly covered with spreading hairs or glands up to 2 mm. long, usually very weak but rather woody at the very base. Leaves generally opposite, sometimes alternate towards the base; lamina 5-lobed, divided to about two-thirds of the radius; lobes 3–10 mm. wide at the base, obovate in outline, pinnatifid, pubescent on both surfaces, usually also hispid with scattered hairs on the upper surface; petiole 0·5–9 cm. long, longitudinally grooved, covered with spreading hairs or glands; stipules divided nearly to the base into lanceolate segments, 2–4 mm. long, pubescent or pilose. Inflorescence 2-flowered; peduncle up to 5 cm. long, but frequently very short or lacking, the flowers then solitary, glandular. Flowers mostly chasmogamous but those on lower stem portion usually cleistogamous; pedicels 2–20 mm. long, pubescent and sometimes glandular; bracts linear-lanceolate, 2–4 mm. long, hyaline at the edge, pubescent. Sepals lanceolate, 4–5·5 mm. long, 1–3 mm. wide, mucronate, 3-nerved, hyaline at the edge, pubescent to pilose or glandular. Petals obovate to broadly spathulate, 6–7·8 mm. long, 3·5–5 mm. wide, obtuse, purple-pink with distinct purple centre. Stamens: filaments narrowly lanceolate or acicular, 2–3 mm. long, purple; anthers 0·6–0·9 mm. long, 0·3–0·5 mm. wide, purple. Ovary pubescent; stigmas 0·3–1 mm. long, dark purple. Mericarps with shallow reticulate ridges, glabrous or shortly pubescent; mature rostrum 8–14 mm. long, pubescent; upper portion of the style 0–2 mm. long. Seeds 2–2·5 mm. long, 1–1·5 mm. wide, glabrous; testa minutely foveolate, pubescent along the minute ridges, with sharp beak, brown to almost black.

UGANDA. Karamoja District: Mt. Moroto, Oct. 1958, *J. Wilson* 562!; Kigezi District: Mabungo to Lake Chahafi, 12 Jan. 1933, *C. G. Rogers* 349!; Busoga District: Bukedi Butandiga, 20 Oct. 1916, *Snowden* 425!
KENYA. Uasin Gishu District: Kipkarren, Dec. 1931, *Brodhurst-Hill* 613!; Kiambu District: Muguga, July 1952, *Verdcourt* 679!; Kisumu–Londiani District: Londiani to Fort Ternan road, shoulder of Limutit, 26 Sept. 1953, *Drummond & Hemsley* 4468!

TANGANYIKA. Moshi District: Lyamungu, 20 Aug. 1932, *Greenway* 3061 !; Pare District: Suji, Aug. 1928, *Haarer* 1487 !; Morogoro District: Nguru Mts., 15 July 1935, *Rounce* 430 !

DISTR. U1–3; K3–6; T1–3, 6–8; Cameroon highlands and highlands of eastern Africa from Eritrea to Rhodesia, also Yemen, Iran, China and the Himalayas

HAB. Upland grassland, wooded grassland, evergreen bushland and thicket, often in densely shaded places, becoming a weed of cultivation; 1000–2900 m.

SYN. *G. favosum* A. Rich. var. *sublaeve* Oliv., F.T.A. 1: 292 (1868). Type: Cameroon Mt., *Mann* 1261 (K, holo. !)
    G. ocellatum Cambess. var. *africanum* Knuth in E.P. IV. 129: 62 (1912), pro parte; Petit in F.C.B. 7: 22 (1958). Many syntypes from eastern Africa, Arabia and Iran
    G. ocellatum Cambess. var. *camerunense* Knuth in E.P. IV. 129: 63 (1912). Type: E. Cameroun Republic, Buea, *Preuss* 983 (BM, iso. !)
    G. brevipes Hutch. & Dalz., F.W.T.A. 1: 138 (1927). Type: as for *G. favosum* var. *sublaeve*
    G. ocellatum Cambess. var. *sublaeve* (Oliv.) Milne-Redh. in K.B. 1948: 453 (1948)
    [*G. simense* sensu Suesseng. in Proc. & Trans. Rhod. Sci. Ass. 43: 108 (1951), *non* A. Rich.]

5. **G. elamellatum** *Kokwaro* in K.B. 23: 527, fig. 1 (1969). Type: Kenya, Naivasha District, Kinangop, *Albrechtsen* 2763 (K, holo. !, BR, EA, iso. !)

Annual herb, slightly viscid, decumbent or ascending, up to 7 dm. long; stems terete or longitudinally grooved; internodes weak and hollow, 3–17·5 cm. long, up to 3·5 mm. thick, mostly pilose below but becoming pubescent with glandular hairs on the younger parts towards the apex. Leaves mostly opposite, sometimes alternate towards the base; lamina 3–5-lobed, pentagonal in outline, divided to the midrib; lobes pinnatisect to palmate-compound, the central one in particular long-attenuate to the base and averaging only 1 mm. wide on this part, hispid on both surfaces; petiole (2–)4–19 cm. long, longitudinally grooved, pubescent and glandular; stipules ovate-lanceolate, 0·3–4 mm. long (much reduced on the older stem nodes), covered with glandular hairs especially on the abaxial surface. Inflorescence 2-flowered; peduncle 1–7·5 cm. long, pilose to glandular hairy; pedicels 4–15 mm. long; bracts often less than 2 mm. long. Sepals ovate, 4–8 mm. long, 2–3 mm. wide, mucronate, 3-nerved, hyaline along the edges, pilose or pubescent and glandular on the outer surface, glandular on the inner surface. Petals oblanceolate-spathulate, 5·5–9 mm. long, 0·5–2 mm. wide, pale lilac-mauve to bright pink, base of the claw white. Stamens: filaments narrowly lanceolate or acicular (truncate at the base), 4–6·5 mm. long; anthers somewhat glandular, 0·2–0·3 mm. long and wide, yellow. Ovary grooved and pubescent along the longitudinal ridges; stigmas 0·5–1 mm. long, light brown to pale cream. Mericarps with shallow reticulate ridges, 2·2–3 mm. long, frequently with raphe at the anterior end, glabrous or occasionally shortly pubescent; mature rostrum 10–19 mm. long, pubescent. Seeds 2–2·5 mm. long, 1–1·5 mm. wide; testa glabrous, pale brown. Fig. 1.

KENYA. Nakuru District: Ol Joro Orok, Jan. 1932, *Pierce in Napier* 1682 !; Naivasha District: Kinangop area, 22 Dec. 1930, *Napier* 729 !; N. Nyeri District: Nanyuki, Jan. 1931, *Hill in F.D.* 2526 !

TANGANYIKA. Mbeya Range, N. aspect, 16 Mar. 1960, *Kerfoot* 1652 !

DISTR. K3–5; T7; Eritrea, Congo and Sudan Republics

HAB. Upland rain-forest and riverine forest, along edges, beside paths and streams and in other moist shady places; 1300–3000 m.

SYN. [*G. purpureum* sensu Milne-Redh. in K.B. 5: 340 (1951); E.P.A.: 347 (1956); Petit in F.C.B. 7: 21 (1958), *non* Vill.]

NOTE. This species is closely related to *G. purpureum* Vill., but differs entirely by the type of ridges on the surface of the mericarps.

6. **G. kilimandscharicum** *Engl.*, Hochgebirgsfl. Trop. Afr.: 274 (1892); R. Knuth in E.P. IV. 129: 157, fig. 20 (1912); A.V.P.: 124 & 293 (1957).

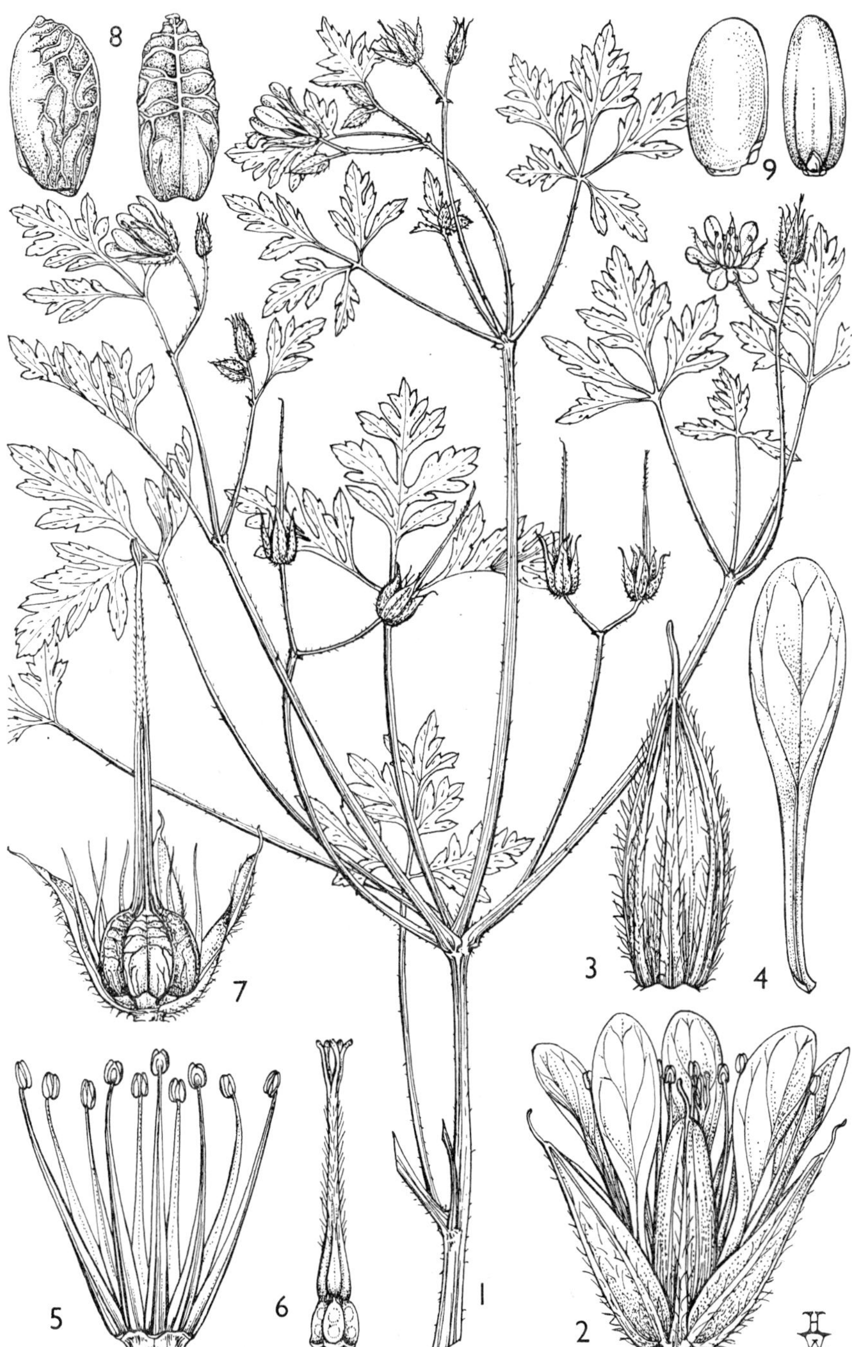

FIG. 1. *GERANIUM ELAMELLATUM*—**1**, habit, × 1; **2**, flower, × 5; **3**, sepal, × 7½; **4**, petal, × 7½; **5**, stamens, × 7½; **6**, gynoecium, × 7½; **7**, fruit, with two sepals removed to show mericarps, × 3; **8**, mericarp, side and dorsal views, × 7½; **9**, seed, side and ventral views, × 7½. 1–6, from *Kerfoot* 1652; 7–9, from *Albrechtsen* in *Napier* 2763.

Types: Tanganyika, Kilimanjaro, upper grassy slopes, *Meyer* 16 & 255 (B, syn. †) & Kilimanjaro, near Peter's Hut, *Hedberg* 1221 (UPS, neo.!, EA, K, S, iso.!)

Perennial prostrate stoloniferous herb up to 4 dm. long; stems frequently tufted, sometimes creeping or scrambling, longitudinally grooved, pilose or hispid, the main stem frequently woody below; prostrate branches sometimes rooting at the nodes. Leaves: radical ones sometimes in basal rosettes (especially in dwarf plants), the cauline ones generally opposite; lamina 5-lobed, reniform in outline, divided up to about two-thirds of the radius, frequently pink-red on the lower surface; lobes 3–10 mm. wide at the base, elliptic to rhombic in outline, pinnatilobed into 2–4(–5) parts but mostly 3-partite and inverted triangular in outline, pilose on both surfaces but rather more pronounced on the veins beneath; petiole 2–18 cm. long, pilose to more pubescent towards the lamina; stipules broadly ovate, 5–11 mm. long, 2–5 mm. wide. Inflorescence of generally solitary flowers in the axils of the smaller of the 2 apparently opposite leaves; peduncle plus pedicel 1–8·5 cm. long, pilose to pubescent; pedicel rather indistinct (flower being solitary) but from the bracts upwards 0·5–3 cm. long; bracts generally 2, lanceolate, each 2–4 mm. long, hyaline along the margin. Sepals broadly lanceolate, 5–9 mm. long, 2–3 mm. wide, cuspidate or barely mucronate, 3-nerved, pilose to pubescent on the outer surface, glabrous inside. Petals obovate to broadly spathulate, 5–9 mm. long, 3–5 mm. wide, obtuse or retuse, pink-red (crimson) to light pink towards the base and with conspicuous red veins. Stamens: filaments lanceolate, 2·5–4 mm. long, light purple or cream, hispid along the margin; anthers 0·6–0·8 mm. long, 0·3–0·5 mm. wide, brown. Ovary pubescent; style rather indistinct, ± 0·5 mm. long, pilose, pink-yellow; stigmas 0·7–1·2 mm. long, usually recurved outward, pink-red to purple, glabrous. Mericarps 5, 2·8–3·5 mm. long, light green, smooth, pilose; mature rostrum 8–12 mm. long, 1·5–2·5 mm. wide, pilose. Seeds ellipsoid, 2–2·3 mm. long, 1–1·3 mm. wide, dark brown; testa glabrescent.

UGANDA. Elgon, Mudangi [Mudange], Aug. 1934, *Synge* 976! & Sasa Camp, 16 Apr. 1950, *Forbes* 271!

KENYA. Mt. Kenya, above Urumandi, 15 Aug. 1944, *Le Pelley* in *Bally* 3472! & western alpine region in Hausburg valley, 9 Aug. 1948, *Hedberg* 1863!

TANGANYIKA. Kilimanjaro, near Peter's Hut, 23 Feb. 1934, *Greenway* 3765! & 16 June 1948, *Hedberg* 1221!

DISTR. **U**3; **K**3, 4; **T**2, 7; not known elsewhere

HAB. Upland moor, on rocky and dry ground with open vegetation, near swamps and along stream banks, less commonly in upland grassland and upland rain-forest; 3200–4400 m. (except specimens from **T**7 which have a lower altitude range, 2100–2400 m.)

VARIATION. *G. kilimandscharicum* in its typical form occurs at higher altitudes than *G. arabicum* and is usually easily distinguished. Nevertheless the two species are closely related and some introgression may occur, as some specimens collected generally below 3200 m. approach the form of *G. arabicum*, e.g. Kilimanjaro, *Verdcourt* 1234. The two specimens seen from the Southern Highlands of Tanganyika (Elton Plateau, *Richards* 14018 and 18503) are also rather intermediate and at a relatively low altitude; the status of these plants remains rather uncertain at present.

7. **G. arabicum** *Forsk.*, Fl. Aegypt.-Arab.: 124 (1775); R. Knuth in E.P. IV. 129: 220 (1912); Christensen in Dansk Bot. Ark. 4 (3): 23 (1922); Laundon in Bol. Soc. Brot., sér. 2, 35: 59, t. 1 (1961) & in F.Z. 2: 133, t. 20/F (1963). Type: Yemen, Kurma, *Forsskål* 735 (C-Forsskål Herb., lecto., BM, photo.!)

Perennial diffusely branched or ascending, occasionally tufted, herb up to 9 dm. long; stems somewhat woody, stoloniferous, creeping and scrambling, longitudinally grooved, pilose or pubescent with short reflexed appressed

hairs at least above, often glabrescent below; internodes (1–)3–25 cm. long,
0·8–2·5 mm. thick. Leaves usually opposite but sometimes in basal rosettes
especially the radical ones, occasionally pinkish-red on the lower surface;
lamina 5-lobed, divided to three-quarters of the radius; lobes 3–10 mm. wide
at the base, broadly elliptic to rhombic in outline, pinnatifid or incised, stiffly
pubescent or pilose on both surfaces especially along the veins; petiole
(1–)4–26 cm. long, pilose; stipules entire, ovate-lanceolate, 5–12 mm. long,
3–5 mm. wide, acute to acuminate, membranous, pilose or pubescent.
Inflorescence usually 2-flowered, the flowers rarely solitary by reduction;
peduncles 3–10 cm. long, grooved or terete, pubescent or pilose; pedicels
1–4 cm. long; bracts lanceolate, 3–5 mm. long, hyaline along the edges. Sepals
broadly lanceolate, 6–10 mm. long, 2–3 mm. wide, cuspidate or mucronate,
3-nerved, pilose to pubescent on the outer surface, glabrous inside. Petals
obovate or spathulate, 6–10 mm. long, 4–6 mm. wide, obtuse or retuse, white
or purple (rarely pale purple) with conspicuous red veins. Stamens: filaments
narrowly lanceolate, 3·5–6 mm. long, cream, hispid along the margins towards
the base; anthers 0·7–1·3 mm. long, 0·5–0·7 mm. wide, brown. Ovary
pubescent; stigmas 1–3 mm. long, purple to dark red. Mericarps smooth,
pilose; mature rostrum 12–20 mm. long, ± 2 mm. wide, pubescent. Seeds
ovoid, 1·9–2·4 mm. long, 1–1·5 mm. wide, glabrous; testa covered with very
shallowly reticulate pattern, dark grey to somewhat brown.

subsp. **arabicum**

Leaf-lamina divided for about three-quarters of the radius, each lobe rhombic in outline and incised.

UGANDA. Karamoja District: Mt. Moroto, Oct. 1958, *J. Wilson* 562A!; Ruwenzori,
Bujuku valley near Bigo Camp, 1 Apr. 1948, *Hedberg* 633!; Elgon, Bulambuli,
12 Nov. 1933, *Tothill* 2377!
KENYA. W. Suk District: Kapenguria, 15 May 1932, *Napier* 1972!; Kiambu District:
South Kinangop, near Sasamua Dam, 13 June 1962, *Coe* 703!; Machakos District:
Chyulu Hills, 2 Apr. 1938, *Bally* 7911!
TANGANYIKA. Kilimanjaro, above Marangu, 21 June 1948, *Hedberg* 1333!; Morogoro
District: Mgeta, Kibuko, Mar. 1955, *Semsei* 2044!; Njombe District: escarpment
W. of Itulo farm, 12 Nov. 1966, *Gillett* 17771!
DISTR. **U**1–3; **K**1–6; **T**1–4, 6–8; widespread in the tropical African highlands, north to
Egypt and the Yemen, also in Madagascar
HAB. Upland rain-forest, bushland, grassland and moor, often in moist places by
streams or in marshes; 1100–3950 m.

SYN. *G. simense* A. Rich., Tent. Fl. Abyss. 1: 116 (1847); Knuth in E.P. IV. 129: 203,
fig. 26/A–B (1912); Milne-Redh. in Mem. N.Y. Bot. Gard. 8: 230 (1953) & in
F.W.T.A., ed. 2, 1: 157, fig. 58 (1954); A.V.P.: 125 & 293 (1957); Petit in
F.C.B. 7: 24 (1958). Type: Ethiopia, Simen, Mt. Silke, *Schimper* II 670
(P, holo.! BM, K, S, iso.!)
G. simense A. Rich. var. *glabrius* Oliv., F.T.A. 1: 291 (1868). Type: Cameroon
Mt., *Mann* 1323 & 1966 (both K, syn.!)
G. simense A. Rich. var. *repens* Oliv., F.T.A. 1: 291 (1868). Type: Fernando Po,
Mann 619 (K, holo.!)
? *G. simense* A. Rich. var. *meyeri* Engl., Hochgebirgsfl. Trop. Afr.: 274 (1892).
Type: Tanganyika, Kilimanjaro, *Meyer* 15 (B, holo. †)
G. simense A. Rich. forma *aprica* Engl., Hochgebirgsfl. Trop. Afr.: 274 (1892).
Types: Ethiopia, *Schimper* 734 (BM, K, isosyn.!) & *Steudner* 980 & Tanganyika,
Kilimanjaro, *von Höhnel* 134 (both B, syn. †)
G. simense A. Rich. forma *umbrosa* Engl., Hochgebirgsfl. Trop. Afr.: 274 (1892),
nom. illegit. Type: as for *G. simense*
G. keniense Standl. in Smiths. Misc. Coll. 68 (5): 7 (1917). Type: Mt. Kenya,
western slopes, *Mearns* 1513 (BM, iso.!)

subsp. **latistipulatum** (*A. Rich.*) *Kokwaro* in K.B. 23: 530 (1969). Type: Ethiopia,
Simen Province, Enschedcap and on Mt. Buahit, *Schimper* II 1378 (P, holo.!, BM, K,
iso.!)

Leaf-lamina deeply palmatifid (dissected almost to the base), each lobe deeply
pinnatifid; stipules broadly ovate.

Kenya. Uasin Gishu District: Mau Range about 10 km. N. of Timboroa, 2 June 1948, *Hedberg* 1082!; Naivasha District: western foot of Aberdares, 22 Sept. 1916, *Dowson* 639!; Masai District: Ngong, Dec. 1930, *Napier* 727!

Tanganyika. Masai District: Rift Valley, Oct. 1925, *Haarer* 132B! & W. side of Lemagrut, Lerong, 1 Feb. 1961, *Newbould* 5637!; Mbulu District: Oldeani, Jan. 1935 *Moreau* 13!

Distr. **K**3, 4, 6; **T**2; Ethiopia

Hab. Upland grassland and bushland, along stream banks or in swampy areas; 1000–2800 m.

Syn. *G. latistipulatum* A. Rich., Tent. Fl. Abyss. 1: 117 (1847)

Note. There do not seem to be enough characters to qualify *G. latistipulatum* to full specific status. Its closer relation with *G. arabicum* is evident, and in my opinion it can be best treated as a subspecies of the latter.

2. **MONSONIA**

L., Syst. Nat., ed. 12, 2: 508 (1767) & Mant. Pl.: 14 (1767)

Annual or perennial herbs (sometimes with annual shoots arising from a woody basal region), or suffrutices, erect or decumbent, rarely acaulescent, variously pubescent, usually glandular. Leaves opposite or alternate, petiolate, serrate, dentate or crenate, sometimes lobed or dissected; stipules filiform to subulate or rarely spinescent. Inflorescence usually a 2–several-flowered simple umbel, occasionally the flowers solitary. Flowers actinomorphic, 5-merous; pedicels often sharply bent when in fruit, bracteate. Sepals imbricate, apiculate with membranous margins. Petals obovate, often with truncate or lobed apex. Stamens 15, all fertile, connate at the base, in 5 bundles of 3 filaments each, the central filament of each triad being longer. Extrastaminal glands 5, adnate to the base of the longer filaments. Ovary 5-lobed and 5-locular, rostrate; loculi 2-ovulate; style absent or if present rather indistinct; stigmas 5, clavate or filiform. Fruit a rostrate schizocarp; mericarps rostrate, 1-seeded, oblique, truncate at the apex of the basal portion and tapering towards the base, hirsute; rostrum persistent, helically twisting when ripe, plumose on the inside. Seeds ± oblong-obovoid, exalbuminous; embryo curved.

About 40 species, mostly native of Africa, SW. Asia and NW. India.

Plants annual; petals 7–15 mm. long:
  Petiole as long as or longer than the lamina;
    lamina ovate in outline .   .   . 1. *M. senegalensis*
  Petiole shorter than the lamina; lamina
    lorate to narrowly oblong in outline   . 2. *M. angustifolia*
Plants perennial; petals (10–)14–28 mm. long:
  Lamina hastate and sometimes 3-lobed in
    outline; anthers 1·6–2 mm. long   . . 3. *M. longipes*
  Lamina lanceolate to ovate in outline; anthers
    2–2·5 mm. long .   .   .   .   . 4. *M. ovata* subsp. *glauca*

1. **M. senegalensis** *Guill. & Perr.* in Fl. Seneg. Tent. 1: 131 (1832); A. Rich., Tent. Fl. Abyss. 1: 115 (1847); F.T.A. 1: 290 (1868); R. Knuth in E.P. IV. 129: 301 (1912); V.E. 3(1): 705 (1915); Burtt Davy, Fl. Pl. & Ferns Transv. 1: 193 (1926); F.P.S. 1: 131 (1950); C.F.A. 1: 258 (1951); F.W.T.A., ed. 2, 1: 157 (1954); Bowden & Müller in F.Z. 2: 139 (1963); Merxmüller & Schreiber, Prodr. Fl. Südwestafr. 64: 5 (1966). Type: Senegal, near Saint-Louis, Lampsar [Lamsar], *Leprieur & Perrottet* (P, holo.!)

Annual prostrate to decumbent or sometimes somewhat erect herb, branching from the often woody base; stems up to 4 dm. long but often shorter (sometimes fruiting at a height of only 7 cm.), frequently with short alternating lateral shoots in one of the axils of the apparently opposite leaves; internodes 0·5–4(–6·5) cm. long; vegetative parts, peduncles and pedicels more or less covered with short patent or somewhat recurved often gland-tipped hairs. Leaves alternate but at every node a leaf is apparently opposite a shoot; lamina ovate in outline, 0·9–4 cm. long, 0·4–2·5(–3·5) cm. wide, acute at the apex, margin serrate or dentate to repand-dentate, the base cordate; petiole 1·2–4·5 cm. long; stipules acicular to subulate, 3–10(–12) mm. long. Inflorescence usually of solitary axillary flowers; peduncle 0·5–2 cm. long, leaf-opposed or in the axil of the smaller of the 2 apparently opposite leaves, often on stunted axillary branches; pedicel 1·8–4 cm. long; bracts 1–2 per pedicel, linear, 5–15 mm. long, 0·5–1 mm. wide. Sepals narrowly oblong-elliptic, with an acicular 1–2 mm. long apical appendage, 6–12 mm. long, 1·5–3·2 mm. wide, densely silky pubescent with long glandless patent hairs mixed with the glandular ones. Petals obovate, 9–12 mm. long, 5·5–7·8 mm. wide, the apex truncate or sometimes shallowly lobed, pink with dark veins. Stamens greenish-yellow; filaments 4·8–6·5 mm. long, pubescent; anthers 0·8–1 mm. long, 0·7–0·8 mm. wide. Extrastaminal glands ovate, sometimes with only the lateral rims clearly visible. Ovary densely woolly; stigmas clavate, 0·8–1·2 mm. long, pubescent on the outer surface. Fruit 7–10·5 cm. long; mericarps 8·5–10 mm. long, 2·2–2·5 mm. wide. Seeds 5–5·5 mm. long, 1·5–2·4 mm. wide; testa minutely reticulate, pale brown. Fig. 2, p. 12.

Kenya. Turkana District: Lokitaung, 23 May 1953, *Padwa* 208!; Masai District: Ol Lorgosailie, 20 July 1947, *Bally* 5142! & Magadi–Nairobi road about 40 km. from Magadi, 10 Apr. 1956, *Greenway* 8994!
Tanganyika. Masai District: Moinik Plateau on W. wall of Rift near Lake Natron, 23 July 1962, *Newbould* 6204!; Mbulu District: E. of Lake Eyasi, Mongala, 1 Apr. 1964, *Verdcourt* 4014B!; Iringa District: Mtera, 19 Apr. 1962, *Polhill & Paulo* 2068!
Distr. **K**2, 6, 7; **T**2, 7; from Senegal across to Egypt, Eritrea, the Somali Republic and Ethiopia, in Angola through South West Africa to N. Transvaal, Botswana and Rhodesia, also in Asia from Arabia across to India
Hab. Deciduous bushland and semi-desert grassland, often on rocky ground, on lava outcrops or in dry sandy places; 200–2100 m. (as low as sea-level outside Flora area)

2. **M. angustifolia** *A. Rich.*, Tent. Fl. Abyss. 1: 115 (1847); Bowden & Müller in F.Z. 2: 140 (1963); Merxmüller & Schreiber, Prodr. Fl. Südwestafr. 64: 3 (1966). Types: Ethiopia, Begemeder, near Gafta, *Schimper* II 1222 (P, syn.!, BM, BR, K, S, UPS, isosyn.!) & Shoa, *Petit* (P, syn.!)

Annual prostrate to decumbent or sometimes ± erect herb, branching from the occasionally semi-woody base; stems up to 5 dm. long, but often shorter, frequently with short lateral shoots in the axils; vegetative parts, peduncles, pedicels, and sepals pubescent, rather densely covered with extremely short recurved or patent and often gland-tipped hairs and also with scattered patent much longer ones. Leaves semi-opposite below, opposite towards the apex; lamina lorate to narrowly oblong or narrowly elliptic, 15–35 mm. long, 4–8(–15) mm. wide, obtuse to subacute at the apex, margin irregularly serrate, the base cuneate; petiole 3–15(–20) mm. long; stipules acicular to subulate, 3–7(–10) mm. long. Inflorescence usually of 2–3 flowers or the flowers occasionally solitary; peduncle 2–20(–40) mm. long, leaf-opposed or in the axil of the smaller of the 2 apparently opposite leaves, sometimes on stunted axillary branches; pedicels 14–42 mm. long; bracts 1–5 per pedicel, linear to filiform, 4–8(–14) mm. long. Sepals narrowly oblong to lanceolate with an acicular 1–2 mm. long apical appendage, 6–10 mm. long, 2·5–3·6 mm. wide, darker medially, hyaline and ciliate along the edges. Petals obovate, 7–12(–15) mm. long, 4–6 mm. wide, apex truncate or shallowly lobed, normally mauve to blue but sometimes white or yellow. Stamens: filaments

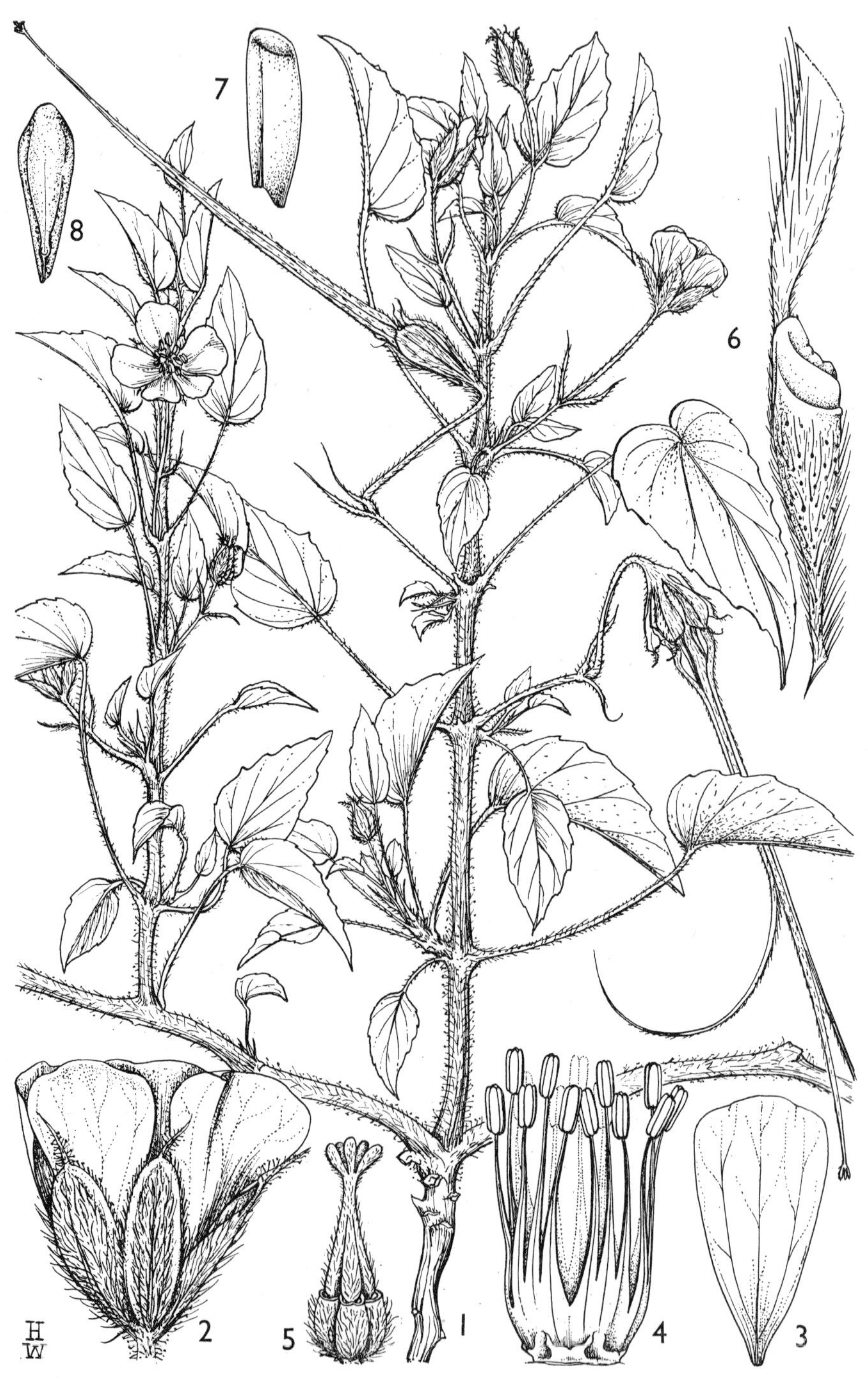

FIG. 2. *MONSONIA SENEGALENSIS*—**1**, habit, × 1; **2**, flower, × 3; **3**, petal, × 3; **4**, stamens, × 5; **5**, gynoecium, × 5; **6**, mericarp, showing coccus and small part of rostrum, × 5; **7**, seed, side view, × 5; **8** same front **view,** × 5.　1–5, from *Bogdan* 3477a; 6–8, from *Bally* 5142.

6–7·2 mm. long, pubescent; anthers 0·8–1·2 mm. long, 0·5–0·8 mm. wide. Extrastaminal glands inconspicuous ovate cavities with sometimes only the 2 lateral rims visible. Ovary tomentose; stigmas clavate, 1·2–2 mm. long. Fruit 5–6·5(–9) cm. long; mericarps 8·5–10 mm. long, 1·7–2·2 mm. wide, pubescent. Seeds 4–5·3 mm. long, 1·3–1·7 mm. wide, generally widest at the apex; testa smooth to minutely reticulate, pale brown to dull yellow.

UGANDA. Karamoja District: E. Matheniko, Sept. 1958, *J. Wilson* 624!; Toro District: Katwe, 24 June 1945, *A. S. Thomas* 4159!; Ankole District: Ruizi R., 10 Nov. 1950, *Jarrett* 178!

KENYA. Northern Frontier Province: Leroghi, Legas Ridge, 17 June 1959, *Kerfoot* 1110!; Kitui District: Migwani–Kakiani about 11 km. N. of Kitui, 4 May 1960, *Napper* 1582!; Masai District: Mara area, Aitong, 5 June 1961, *Glover et al.* 1562!

TANGANYIKA. Musoma District: Serengeti, N. of National Park, 5 May 1958, *Paulo* 432!; Mpwapwa District: Kongwa, Lubiri Mbuga, 28 Feb. 1949, *Anderson* 353!; Mbeya District: S. end of Rukwa Rift, Mjere, 27 Feb. 1934, *Michelmore* 976!

DISTR. **U**1, 2; **K**1, 3–7; **T**1, 2, 4, 5, 7; eastern Africa from the Somali Republic to the Cape Province of South Africa, also Madagascar

HAB. Deciduous woodland and bushland, wooded or open grassland, sometimes on river banks and in wet grasslands, sometimes on alluvial soil or black clay of volcanic origin; 390–2290 m.

SYN. [*M. biflora* sensu Harv. in Harv. & Sond., Fl. Cap. 1: 255 (1860); F.T.A. 1: 290 (1868); Engl., Hochgebirgsfl. Trop. Afr.: 275 (1892); R. Knuth in E.P. IV. 129: 305 (1912), pro parte; E.P.A.: 348 (1956); Petit in F.C.B. 7: 25 (1958), *non* DC.]

    M. biflora DC. var. *angustifolia* (A. Rich.) Burtt Davy, Fl. Pl. & Ferns Transv.: 46 (1926) & 193 (1926), pro parte

3. **M. longipes** *Knuth* in E.J. 40: 66 (1907) & in E.P. IV. 129: 294, fig. 38/B (1912); W.F.K.: 22 (1948); E.P.A.: 348 (1956). Type: Kenya, Machakos District, Makindu, *Kassner* 538 (BM, K, iso.!)

Perennial decumbent herb up to 6 dm. long, generally with a thick rootstock; stems profusely branched from the rootstock and frequently forming a spreading mat, woolly with long as well as short hairs (long hairs sometimes glandular, filled with honey-like juice which gives a creamish pubescence to the whole plant); internodes (0·5–)2–10(–15) cm. long. Leaves mostly opposite, sometimes alternate especially the lower ones, or occasionally in apparent whorls in short-stemmed plants; lamina usually hastate in outline, 1·5–5·5 cm. long, 0·7–2·7 cm. wide, acuminate at the apex, margins dentate or double-serrate to almost entire at the broadly hastate base (the whole margin ciliate under magnification), hispid on the upper surface to tomentose on the lower surface particularly along the veins; petiole 1–7 cm. long, pubescent; stipules acicular, 5–18 mm. long, hairy. Inflorescence of 1–5 flowers; peduncle (3–)6–18 cm. long, leaf-opposed or in the axil of the smaller of the 2 apparently opposite leaves, hispid or pubescent; pedicels 5–20 mm. long, pubescent. Sepals spathulate to oblong and with an acicular 2–4·5 mm. long apical appendage, 10–18 mm. long, 3·5–7 mm. wide, hispid on the outer surface, glabrous inside, ciliate along the margins. Petals obovate to broadly oblanceolate, 15–25 mm. long, 7–11 mm. wide, mucronate to obtuse, bright yellow to cream-lemon (sometimes white). Stamens: filaments 9–12 mm. long, pubescent; anthers 1·6–2 mm. long, 0·6–0·8 mm. wide, yellow. Extrastaminal glands rather inconspicuous. Ovary tomentose; stigmas clavate, 1·5–2·5 mm. long. Fruit 6–9·2 cm. long; mericarps 10–13 mm. long, 1·8–2·3 mm. wide, pubescent. Seeds 4–5·5 mm. long, 1·8–2·5 mm. wide (usually widest at the apex); testa minutely reticulate, pale brown or pinkish-brown.

KENYA. Northern Frontier Province: Marsabit, July 1942, *Bally* 1859!; Machakos District: Emali Station, 30 Jan. 1953, *Bally* 8606!; Masai District: Nairobi–Magadi road about 80 km. from Nairobi, 8 Apr. 1956, *Greenway* 8986!

Tanganyika. Shinyanga, 3 Apr. 1932, *B. D. Burtt* 3741 !; Mbulu District: W. slopes of Mt. Hanang, Kendabi [Gendabi], 10 Feb. 1946, *Greenway* 7689 !; Mbeya District: Usangu Plain near Utengule, 28 Jan. 1963, *Richards* 17590 !

Distr. **K**1, 3, 4, 6; **T**1, 2, 7; Ethiopia

Hab. In grassland and bushland, in semi-seasonal swamps and along river banks, on sandy (mostly lateritic) loam; 750–2500 m.

4. **M. ovata** *Cav.*, Diss. 4: 193, t. 113/1 (1787); Harv. in Harv. & Sond., Fl. Cap. 1: 255 (1860); R. Knuth in E.P. IV. 129: 302 (1912); Bowden & Müller in F.Z. 2: 137 (1963). Types: South Africa, Cape of Good Hope, *Thunberg* (MA, holo., UPS, Thunberg Herb. Nos. 15783 & 15784, iso.!)

Perennial erect herb up to 4 dm. tall, with ± woody basal region and thick fleshy roots; stems woody with annual leafy branches; internodes 0·1–3(–5) cm. long; vegetative parts, peduncles, pedicels, and sepals pubescent, ± covered with short recurved or patent and often gland-tipped hairs and with long patent hairs also present. Leaves opposite to subopposite (almost alternate), or just in a bunch of indistinct phyllotaxy especially on the annual shoots; lamina narrowly lanceolate to ovate, 1–4·5 cm. long, 0·3–2·3 cm. wide, acute to subacute at the apex, margin serrate to crenate, the base cordate to subcordate, tomentose; petiole 1–2 cm. long, densely pubescent with short and long as well as glandular hairs; stipules acicular to spiny, 5–12(–14) mm. long. Inflorescence usually of 2 flowers; peduncles 1·5–9 cm. long, leaf-opposed or in the axil of the smaller of the 2 apparently opposite leaves; pedicel (1–)2·5–5(–7) cm. long; bracts 2–6 per pedicel, filiform to linear, 5–12 mm. long. Sepals narrowly oblong to lanceolate, with an acicular 1–5 mm. long apical appendage, 7–12 mm. long, 3–4 mm. wide. Petals obovate, (11–)14–22·5 mm. long, 6–14(–17) mm. wide, truncate at the apex, white, cream or yellowish. Stamens yellow; filaments 7–9·5 mm. long, pubescent; anthers 2–2·5 mm. long, 1–1·5 mm. wide. Extrastaminal glands inconspicuous ovate cavities with sometimes only the 2 lateral rims visible. Ovary tomentose; stigmas clavate, 1·5–2·5 mm. long. Fruit 5·5–10 cm. long; mericarps 12·5–13·5 mm. long, 2–2·6 mm. wide. Seeds 6–6·5 mm. long, 2–2·4 mm. wide; testa minutely reticulate, pale brown.

subsp. **glauca** (*Knuth*) *Bowden & T. Müller* in F.Z. 2: 137 (1963); Merxmüller & Schreiber, Prodr. Fl. Südwestafr. 64: 4 (1966). Type: Tanganyika, Kilimanjaro, Marangu, *Volkens* 2128 (BM, K, iso. !)

Leaf-lamina 2·0–4·5 cm. long, 4–22 mm. wide; petiole 1–4·5(–5) cm. long; sepals with apical appendage 1–2·5 mm. long; fruit 6·5–10 cm. long.

Kenya. Machakos District: S. face of Lukenya, 30 May 1958, *Verdcourt & Napper* 2171 ! & Lukenya, 14 May 1961, *J. G. Williams* in *E.A.H.* 12325 !

Tanganyika. Mbulu District: N. Sambala Hills, Lankasese path, 29 Mar. 1929, *B. D. Burtt* 1983 !; Moshi District: Himo R. to Taveta road, 25 Jan. 1936, *Greenway* 4502 !; Pare District: near Same, Oct. 1927, *Haarer* 869.

Distr. **K**4; **T**1–3; Botswana, Rhodesia, South West Africa and South Africa (Transvaal)

Hab. Deciduous bushland or wooded grassland, on sandy and stony or volcanic soils; 800–1500 m.

Syn. *M. glauca* Knuth in E.J. 40: 64 (1907)

Note. Subsp. *ovata* is restricted to the Cape Province of South Africa. The distinguishing features are given by Bowden and Müller in F.Z. 2: 139 (1963).

3. **PELARGONIUM**

Ait., Hort. Kew. 2: 417 (1789)

Annual or perennial herbs, shrublets or shrubs, sometimes with underground tubers, erect or decumbent, sometimes acaulescent or with sub-succulent stems; stems viscid and aromatic, variously hairy, often glabrous.

Leaves opposite or alternate, usually petiolate, serrate, crenate, dentate, lobed, variously dissected or compound, rarely entire; stipules membranous, sometimes setaceous, rarely forming spines. Inflorescence a 2–many-flowered simple umbel, flowers rarely solitary. Flowers zygomorphic, mostly 5-merous, pedicellate, bracteate. Sepals imbricate, often with a membranous margin; posterior sepals produced at the base into a spur which is decurrent along the pedicel and adnate to it. Petals 5, rarely 0–4 by abortion, usually unequal, imbricate, unguiculate or sessile. Stamens 10, connate at the base, with only 2–7 filaments bearing anthers, the remaining ones vestigial (staminodes); anthers dorsifixed. Extrastaminal glands absent. Ovary 5-lobed, 5-locular (occasionally 3–4-locular by abortion), rostrate; each locule usually 2-ovulate, 1 ovule later aborting; style short or long; stigmas 5, generally filiform. Fruit a rostrate schizocarp; mericarps rostrate, 1-seeded, tapering from the apex to the base and ending in a point, hirsute; rostrum persistent, helically twisting when ripe, plumose on the inside. Seeds ± oblong-obovoid, exalbuminous; embryo curved.

About 300 species, mostly of tropical and subtropical regions, especially numerous in South Africa.

A good number of them are commonly grown as ornamentals and yield the well known " geraniums " of the florists. Some of these may become naturalized in East Africa; *P. inquinans* (L.) Ait. has been found in natural vegetation N. of Nyeri (*Verdcourt* 3458 !) and *P. papilionaceum* (L.) Ait. has been collected from half way up Ruwenzori (*Mason* in *Herb. Hort. Kew.* H 925/65 !). See also p. 1.

Plant acaulescent . . . . . . 1. *P. luridum*
Plant with short or long stem :
  Leaves semi-compound, 3– or more partite to
    foliolate . . . . 2. *P. whytei*
  Leaves simple, the lamina almost entire to
    palmately 3–5(–7)-lobed :
    Petals present :
      Number of petals 4 . . . . . 3. *P. glechomoïdes*
      Number of petals 5 :
        Petals greenish-yellow; lamina distinctly
          3–5-partite, hastate in outline . . 4. *P. quinquelobatum*
        Petals white, pink or dark red; lamina
          palmatilobed, cordate in outline . 5. *P. alchemilloïdes*
    Petals absent . . . . . . 6. *P. apetalum*

1. **P. luridum** (*Andr.*) *Sweet* in Colv. Cat., ed. 2: 22 (?1822) & Geran. 3, t. 281 (1825); C.F.A. 1: 259 (1951); D. Clifford, Pelargoniums: 173 (1958); Petit in F.C.B. 7: 26 (1958); Müller in F.Z. 2: 141 (1963). Type: plate 34 in H.C. Andrews, Geraniums 2 (?1813), drawn from a plant cultivated at Hammersmith Nursery (London), seeds originally from South Africa

Perennial erect herb up to 7 dm. tall, with a tuberous woody rootstock, acaulescent; vegetative parts, peduncle and pedicels glandular and pubescent with both long (± 2 mm.) patent and much shorter somewhat appressed hairs, as well as sessile glands. Leaves mostly developing after the flowering period, all radical; lamina ovate to broadly ovate in outline, (3–)7–15 (–24) cm. long, 4–14(–19) cm. wide, cuneate to cordate at the base, extremely variously dissected, shallowly lobed to pinnatisect or bipinnatisect, later leaves often more dissected; ultimate segments filiform to oblong or ovate, entire to serrate or crenate towards the apex, when entire apex acute to rounded or mucronate, densely tomentose to ± glabrous, with long patent hairs absent or sparse on many leaves and always restricted to the margins and veins; petiole 6–20(–30) cm. long; stipules linear to narrowly triangular,

15–40 mm. long, 1·5–5 mm. wide, acuminate. Inflorescence a terminal simple umbel of 5–30(–50) flowers; peduncles 1–3 per plant, each 14–65 cm. long; free part of the pedicel 2–35(–50) mm. long; bracts few to numerous per pedicel, linear to lorate or lanceolate, 8–16 mm. long, 1·5–4 mm. wide, acuminate, membranous. Sepals lanceolate to narrowly ovate, 8·5–15 mm. long, 1·5–5 mm. wide, glandular and densely pubescent. Spur 3–7 cm. long. Petals 5, oblanceolate to narrowly obovate, 12–24(–28) mm. long, 5–12 mm. wide, white to pale yellow with pink venation or purple, the 3 anterior ones clawed. Stamens connate at the base for 1·8–3·5 mm.; antheriferous filaments usually 7, 4–8 mm. long; staminodes 3, 3–5·5 mm. long; anthers 1·6–3·2 mm. long, 1·2–1·7 mm. wide. Ovary tomentose at the base; rostrum pubescent; style 0·1–1·2 mm. long; stigmas 1·6–2·8 mm. long. Fruit 4–5(–6·5) cm. long; mericarps 9–15 mm. long, 1·5–3 mm. wide. Seeds 4·8–6 mm. long, 1·9–2·2 mm. wide; testa minutely reticulate, pale brown.

Tanganyika. Ufipa District: Mbisi [Mbizi] Forest, 26 Nov. 1958, *Napper* 1025!; Mbeya District: Mbozi, 28 Aug. 1933, *Greenway* 3614!; Songea District: Matengo Hills, Lupembe ridge near Mpapa, Nov. 1951, *Eggeling* 6365!

Distr. **T**4, 7, 8; Angola and the Congo Republic southwards to the north-eastern parts of South Africa

Hab. Upland grassland and grassy places in deciduous woodland or in secondary bushland, often in places subject to fire, sometimes by streams or on damp ground; 800–2440 m.

Syn. *Geranium luridum* Andr., Geran. 2, t. 34 (?1813)
    Polyactium aconitiphyllum Eckl. & Zeyh., Enum. Pl. Afr. Austr. Extratrop.: 67 (1835). Type: South Africa, Cape Province, near Kei R., *Ecklon & Zeyher* (B, holo. †)
    Pelargonium aconitiphyllum (Eckl. & Zeyh.) Steud., Nom. Bot. 2: 283 (1841); R. Knuth in E.P. IV. 129: 361 (1912)
    P. flabellifolium Harv. var. *benguellense* Oliv., F.T.A. 1: 294 (1868). Type: Angola, Huila, between Lopolo and Marimo, *Welwitsch* 1605 (BM, iso.!)
    P. heckmannianum Engl. in E.J. 30: 335 (1901); R. Knuth in E.P. IV. 129: 363 (1912). Type: Tanganyika, Njombe District, Ukinga Mts., Bulongwa Mt., *Goetze* 1224 (B, holo. †, BR, iso.!)
    P. benguellense (Oliv.) Engl. in Warb., Kunene-Samb.-Exped. Baum: 258 (1903)

Note. Fuller synonymy of this very variable species, particularly that referring to southern Africa, is given by Müller in F.Z. 2: 141 (1963).

2. **P. whytei** *Bak.* in K.B. 1897: 246 (1897); R. Knuth in E.P. IV. 129: 394 (1912); V.E. 3(1): 711 (1915); Brenan in Mem. N.Y. Bot. Gard. 8: 231 (1953); A.V.P.: 125 (1957); D. Clifford, Pelargoniums: 146 (1958); Petit in F.C.B. 7: 28, t. 2 (1958); Müller in F.Z. 2: 143 (1963). Type: Malawi, Nyika Plateau, *Whyte* 244 (K, holo.!)

Perennial decumbent (occasionally tufted) herb; stems up to 1·3 m. long, straggling, often reddish to pink, branching, subsucculent, arising from a woody base; internodes up to 13 cm. long; vegetative parts, peduncles and pedicels glandular, ± pubescent to almost glabrous, hairs rather patent or somewhat adpressed on the leaves; glands sessile, glittering-gold; glandular hairs on the red stem parts usually filled with red pigments. Leaves opposite, sometimes alternate towards the stem-base; lamina narrowly ovate to ovate in outline, 1·5–5·5 cm. long, 1·5–5 cm. wide, cordate, 3-partite to 3-foliolate (sometimes much dissected, giving several subopposite pairs of pinnae plus the terminal one), with a larger middle section of the leaflet, sometimes only shallowly 3-lobed or pinnatifid to pinnatisect; segments or pinnae somewhat rhombic in outline, almost unlobed to pinnatipartite, the margin crenate, the hairs scattered and often restricted to the margins and veins; petiole 2–8(–12) cm. long; stipules narrowly to broadly ovate, 5–10 mm. long, 2–7 mm. wide, acute to apiculate at the apex, membranous. Inflorescence a terminal simple umbel of 2–4(–5) flowers or the flowers sometimes solitary; peduncle 5–9(–13) cm. long, leaf-opposed or in the axil of the smaller of the 2 apparently opposite leaves; pedicels 1–3 mm. long; bracts 4–8 per node, lanceolate, acute, membranous, 5–8 mm. long, 1·5–3 mm. wide. Sepals lanceolate, 9–14 mm. long, 1–4 mm. wide, acuminate, glandular and sparsely pubescent, with conspicuous red veins. Spur 5–15(–20) mm. long. Petals 4, oblanceolate and unguiculate to ± spathulate, pink with red veins, the 2 posterior ones 9–20 mm. long, 3–5·5 mm. wide, generally longer than the anterior ones; the anterior ones 9–15 mm. long, 2–4 mm. wide. Stamens connate at the base for 1–2·5(–5) mm.; antheriferous filaments 7, 5–12 (–16) mm. long, white below and pink towards the tip; staminodes 3, 2·5–6·5 mm. long, mostly white throughout; anthers 2–2·5 mm. long, ± 1 mm. wide. Ovary with basal part tomentose; rostrum pubescent; style 3–8 mm. long; stigmas 1·5–3 mm. long, pink. Fruit 3–4 cm. long; mericarps 5–7 mm. long, 1·5–2 mm. wide. Seeds 3·5–5 mm. long, 1·5–1·7 mm. wide; testa minutely reticulate, pale brown.

UGANDA. Karamoja District: Mt. Moroto, June 1963, *Tweedie* 2659 !; Elgon, Mudangi [Mudange], May 1935, *Synge* 1925 !

KENYA. Naivasha District: top of Longonot Crater, 29 June 1930, *Napier* 213A !; Machakos District: Chyulu North, 21 Apr. 1938, *Bally* 208 !; Masai District: SE. slope of Ngong Range, 29 Dec. 1932, *C. G. Rogers* 260 !

TANGANYIKA. W. Usambara Mts., Mtai–Sunga road escarpment, 25 May 1953, *Drummond & Hemsley* 2760 !; Rungwe District: on slopes of Ngosi [Wentzel Heckmann] Crater, 28 Aug. 1936, *B. D. Burtt* 6258 !; Njombe District: by R. Hagafilo about 11 km. S. of Njombe, 8 July 1956, *Milne-Redhead & Taylor* 11012 !

DISTR. **U**1, 3; **K**1, 3, 4, 6; **T**2, 3, 7; Congo Republic, Ethiopia, Malawi, Zambia

HAB. In drier parts of upland rain-forest, ericaceous moor, upland grassland and evergreen bushland, often on rocky ground; 1200–3500 m.

SYN. *P. goetzeanum* Engl. in E.J. 30 : 334 (1901). Type: Tanganyika, Njombe District, Ukinga Mts., Yawiri Mt., *Goetze* 1189 (B, holo. †)

   P. gallense Chiov. in Ann. Bot. Roma 9 : 318 (1911). Type: Ethiopia, Shoa Province, Mt. Uaciaccia, above the plain of Daleccia, *Negri* 419 (FI, syn. !)

VARIATION. There is some variation in the dissection of the leaf-blade which may call for some kind of infraspecific division, but on present evidence this character seems somewhat inconsistent.

3. **P. glechomoïdes** *A. Rich.*, Tent. Fl. Abyss. 1 : 118 (1847); F.T.A. 1 : 294 (1868); R. Knuth in E.P. IV. 129 : 454 (1912); Hutch. & Bruce in K.B. 1941 : 95 (1941); E.P.A. : 351 (1956); D. Clifford, Pelargoniums : 148 (1958). Type: Ethiopia, Shoa Province, *Petit* (P, holo. !)

Perennial prostrate or ascending delicate herb up to 5 dm. long, rhizomatous; stems rather lax, branching on the upper part, entirely herbaceous; internodes 2–7·5 cm. long, generally 1–2 mm. across (occasionally enlarged up to 5 mm. thick); stems, peduncles and pedicels covered with short glandular hairs intermingled with long scattered spreading hairs. Leaves opposite and/or alternate, sometimes only in basal rosettes, radical ones ultimately deciduous; lamina reniform, slightly 5–7-lobed, 1–5·5 cm. long, (0·5–)2–9 cm. wide, cordate at the base, membranous, pilose to pubescent especially on the nerves beneath to almost glabrous on the upper surface; lobes deeply crenate-dentate; petiole 1–15·5 cm. long, pubescent with long hairs; stipules ovate-lanceolate, 3–6 mm. long, 1·5–3 mm. wide, acuminate, membranous. Inflorescences of 2–8 flowers, borne alternately along the stem; peduncle 2–10 cm. long, leaf-opposed or in the axil of the 2 apparently opposite leaves; free part of the pedicel 5–15 mm. long; bracts lanceolate, 3–5 mm. long, 1–1·5 mm. wide, acute, hyaline and ciliate along the margins, membranous. Sepals lanceolate to ovate, 4–9 mm. long, 1·5–4 mm. wide, usually coloured like the petals, hyaline on the margin, cuspidate to acuminate, pubescent and glandular. Spur funnel-shaped, 5–18 mm. long. Petals 4, 7–15 mm. long, 2–6 mm. wide towards the apex, bright pink or rose-red with crimson lines, the 2 anterior ones spathulate, the posterior pair oblong-cuneate, all much narrowed towards the base and entire. Stamens connate at the base for 1–2·7 mm.; antheriferous filaments 5, 6–10 mm. long; staminodes 5, subulate, 1–3 mm. long, alternating with the antheriferous filaments; anthers 1·5–1·8 mm. long, 0·6–0·8 mm. wide. Ovary with basal part tomentose; stigmas 0·6–0·8 mm. long; style glabrous, 6–8 mm. long. Fruit 1·5–2·5 cm. long, pilose to pubescent; mericarps 4–5·2 mm. long, 2·4–3·2 mm. wide. Seeds oblong-ovoid, 2·2–3 mm. long, 1–1·5 mm. wide, glabrous; testa smooth, brown. Fig. 3.

KENYA. Northern Frontier Province: Mt. Marsabit, July 1942, *Bally* 1860! & Ndoto Mts., June 1937, *Jex-Blake* in *C.M.* 6906!; Machakos District: 8 km. N. of Ulu Station, Mar. 1930, *Napier* 18!
DISTR. **K**1, 4; Ethiopia, Eritrea and the Somali Republic (N.)
HAB. Deciduous bushland and in grassland, on cliffs and lava outcrops, along river banks, in shade and on sandy soils; 750–2700 m.

4. **P. quinquelobatum** *A. Rich.*, Tent. Fl. Abyss. 1 : 118 (1847); F.T.A. 1 : 293 (1868); R. Knuth in E.P. IV. 129 : 432 (1912); E.P.A.: 351 (1956); D. Clifford, Pelargoniums: 147 (1958). Type: Ethiopia, Tigre Province, near Gapdia, *Schimper* II 792 (P, holo.!, BM, K, iso.!)

Perennial decumbent, straggling or semiprostrate herb up to 7·5 dm. long; rootstock usually tuberous, up to 1·4 cm. in diameter; stems slender, herbaceous, hollow, mostly covered with a deflexed pubescence intermingled with long patent hairs, angled and often swollen at the nodes; internodes sometimes much reduced in length. Leaves may be opposite, sometimes alternate especially the lower ones, frequently in basal rosettes; lamina 3- or 5-partite, hastate in outline, 2·5–12·2 cm. long, 2–10·5 cm. wide, long-haired; petiole 1·5–18 cm. long, covered with short intermingled with long hairs; stipules linear-lanceolate, 5–10(–15) mm. long, 2–3 mm. wide at the base, long-haired on the abaxial surface and along the margins, subglabrous on the inner surface, cuspidate (sometimes bifid) at the apex. Inflorescence of (2–)4–9 flowers; peduncles 6–36 cm. long, mostly leaf-opposed, pubescent with short and long hairs; free part of the pedicel 2–6 mm. long; bracts usually more than the number of the flowers, linear-lanceolate, 5–10 mm. long, 1–1·5 mm. wide, covered with few but long hairs. Sepals linear-lanceolate, lorate or narrowly oblong, 6–12 mm. long, 1–3 mm. wide, glandular and rather pubescent, ciliate and hyaline along the margins. Spur 15–35 mm. long, adnate nearly to the base, hispid and glandular. Petals 5, oblong-spathulate, greenish-yellow, creamy-green, pale lemon greenish-white, or just green with purple veining at the base of the posterior petals, the 2 posterior ones 9–16 mm. long, 4–6 mm. wide, the 3 anterior ones 7–12 mm. long, 3·5–5 mm. wide. Stamens connate at the base for 1–2·5 mm., crimson to purple; antheriferous filaments 7, 2–5 mm. long; staminodes 3, 1–1·8 mm. long; anthers 1·1–1·6 mm. long, 0·6–0·8 mm. wide, red. Ovary tomentose (woolly); rostrum indistinct at the flowering stage; style 0·5–1 mm. long; stigmas 1·2–1·6 mm. long, pink-red. Fruit 2·5–5·2 cm. long, at maturity equal or slightly longer than the spur plus pedicel, pubescent; mericarps 6·5–8 mm. long, 1·5–2 mm. wide. Seeds 3·9–4·4 mm. long, 1–1·3 mm. wide; testa minutely reticulate, pale brown.

UGANDA. Karamoja District: Pian County near Mt. Moroto, 15 July 1959, *Kerfoot* 1279! & Lodoketemit [Lodoketeminit], 8 Nov. 1962, *Kerfoot* 4452!
KENYA. Northern Frontier Province: Moyale, 2 July 1952, *Gillett* 13467!; Nairobi District: Mbagathi, W. of Archer's farm, 9 Feb. 1933, *C. G. Rogers* 406!; Teita District: near Taveta, Maktau, 4 May 1950, *Jeffery* 738!
TANGANYIKA. Musoma District: Naabi Hill, 11 Apr. 1961, *Greenway & Turner* 10025!; Lushoto District: Mombo, 1 Aug. 1960, *Semsei* 3069!; Dodoma District: Kazikazi, 20 May 1932, *B. D. Burtt* 3623!

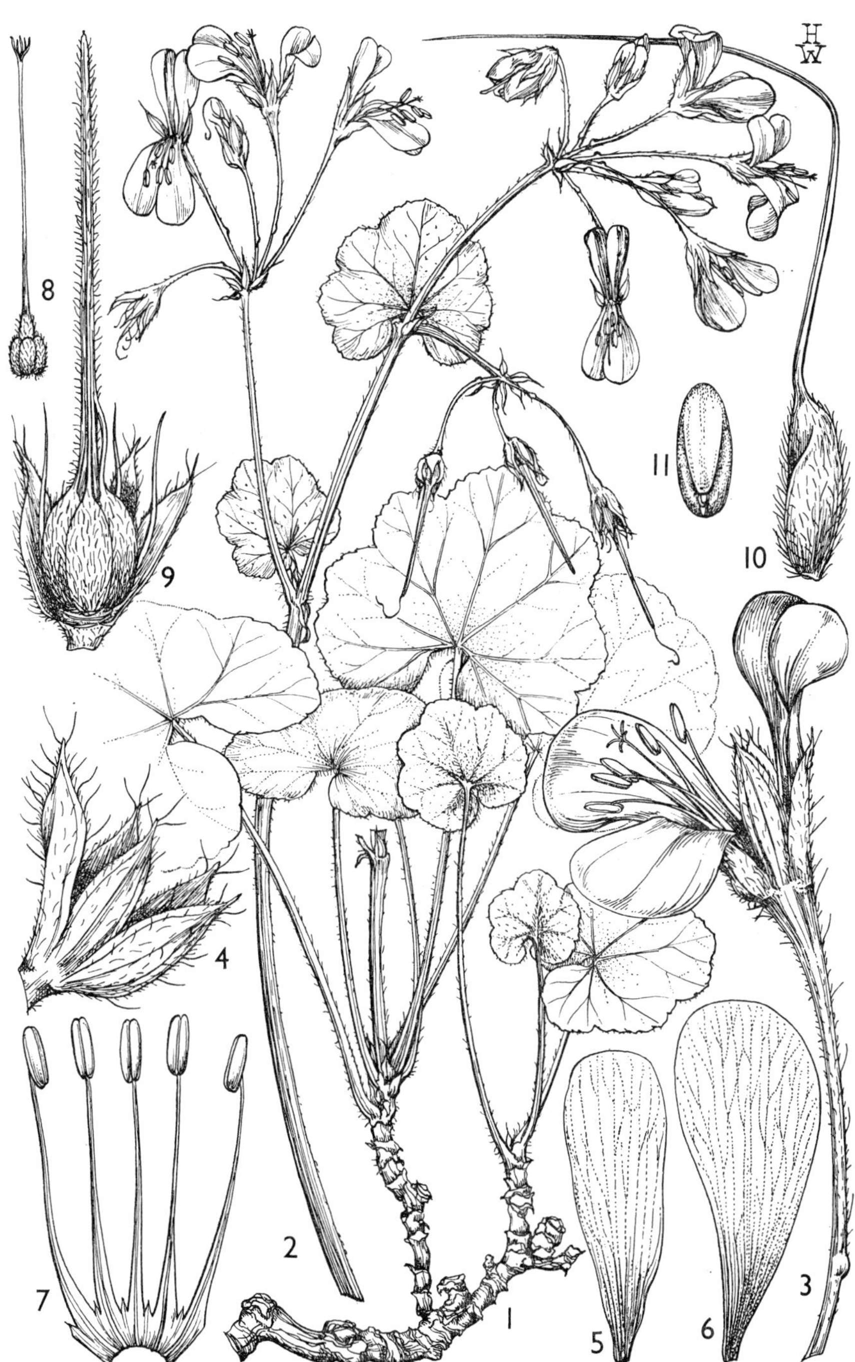

Fig. 3. *PELARGONIUM GLECHOMOIDES*—**1,** base of plant, × 1; **2,** flowering stem, × 1; **3,** flower, × 3; **4,** calyx (without spur), × 4; **5,** posterior petal, × 4; **6,** anterior petal, × 4; **7,** stamens and staminodes, opened out, × 4; **8,** gynoecium, × 4; **9,** fruit, with two sepals removed to show mericarps, × 3; **10,** mericarp, × 5; **11,** seed, × 5.   1, 9–11, from *Burger* 1276; 2, from *Bally* 10084; 3–8, from *Burger* 3575.

Distr. **U**1; **K**1, 3, 4, 6, 7; **T**1–3, 5, 7; Ethiopia and the Somali Republic
Hab. Wooded grassland, bushland and thicket, often on dry rocky slopes, lava outcrops
or on river banks, rarely at forest edges, sometimes becoming a weed of cultivated
ground; 700–2100 m.

Syn.   *Geraniospermum quinquelobatum* (A. Rich.) O. Ktze., Rev. Gen. Pl. 1: 94 (1891)
       Pelargonium fischeri Engl. in Hochgebirgsfl. Trop. Afr.: 276 (1892). Type:
          Tanganyika, Wadiboma (? Handeni District, Kwediboma), *Fischer* 87 (B,
          holo. †)
       P. inaequelobum Mast. in Gard. Chron. 29: 133 (1901). Type: Tanganyika,
          Usagara Mts., *Kirk* (K, holo. !)
       P. somalense Franch. in Révoil, Faune & Flore des Çomalis: 19 (1882). Type:
          Somali Republic, without precise locality, *Révoil* 24 (P, holo. !, K, photo. !)
       P. quinquelobatum A. Rich. var. *migiurtinorum* Chiov., Fl. Som. 1: 122 (1929).
          Type: Somali Republic(S.), Migiurtin Coast, *Stefanini & Puccioni* 780 (FI,
          holo. !)

Variation. There is considerable variation in dissection of the leaf, the plants with the
less dissected (mostly 3-lobed) leaves being predominant in **T**3, 5, 7.

5. **P. alchemilloïdes** (*L.*) *Ait.*, Hort. Kew 2: 419 (1789); Harv. in Harv. &
Sond., Fl. Cap. 1: 295 (1860); R. Knuth in E.P. IV. 129: 428 (1912); V.E.
3(1): 708 (1915); Burtt Davy, Fl. Pl. & Ferns Transv. 1: 190 (1926); D.
Clifford, Pelargoniums: 146 (1958); Müller in F.Z. 2: 146 (1963). Type: a
plant from South Africa cultivated in Europe (Hort. Upsaliensis), Linnean
Herbarium (London) No. 858.24 (LINN, lecto. !)

Perennial decumbent or ascending herb with woody stoloniferous to
tuberous rootstock; stems up to 8 dm. long, slightly branched, hispid to
pubescent and glandular; hairs generally short, patent; glands sessile. Leaves
often alternate on the lower stem portion and opposite towards the apex;
lamina palmatifid to palmatipartite (generally palmatilobed), 2·5–11 cm.
long, 2·5–13 cm. wide, cordate at the base, hispid to pubescent especially
along the veins (hairs ± appressed); lobes 5, rounded to oblong, crenate to
serrate along the margins; petiole (1·5–)3–20 cm. long, pubescent, occasionally with long patent hairs; stipules narrowly to broadly ovate, 5–17·5 mm.
long, 2–17 mm. wide, acute or apiculate to cuspidate, occasionally bifid,
membranous, hispid on the surface to densely ciliate along the margins.
Inflorescence of 2–16 flowers (generally averaging 5–8); peduncle 6–25
(–30) cm. long, leaf-opposed or in the axil of the smaller of the 2 apparently
opposite leaves, glandular and pubescent; free part of the pedicel 3–7
(–11) mm. long, covered with retrose hairs; bracts few to numerous, lanceolate or lorate to narrowly oblong, 3–8·5 mm. long, 0·8–4 mm. wide,
membranous. Sepals lorate to narrowly oblong or lanceolate, 7–10 mm.
long, 1·8–3·5 mm. wide, acute, glandular and ± pubescent (usually with long
scattered hairs), sometimes accrescent. Spur 16–45 mm. long, pubescent
and glandular. Petals 5, usually white but sometimes pink or dark red, the
2 posterior ones (9–)13–18(–20) mm. long, (3–)6–8(–10) mm. wide, the 3
anterior ones oblanceolate or spathulate, clawed, 8–16 mm. long, 3–8 mm.
wide. Stamens connate at the base for 1–2·3 mm.; antheriferous filaments
5–7, 3·5–6·8 mm. long; staminodes 3–5, 2·8–5 mm. long, alternating with the
fertile filaments; anthers 1·4–1·8 mm. long, 0·8–1·2 mm. wide. Ovary
tomentose; rostrum indistinct at the flowering stage; style 0·2–1·5 mm. long;
stigmas 1·2–2·5 mm. long, sometimes divided almost to the base thus leaving
only a short style. Fruit 2·8–4·5 cm. long; mericarps 5, 5·5–7 mm. long,
1·5–2·1 mm. wide. Seeds 4·2–4·6 mm. long, 1·4–1·7 mm. wide; testa minutely
reticulate, pale brown.

Syn.   *Geranium alchemilloïdes* L., Sp. Pl.: 678 (1753)

   subsp. **multibracteatum** (*A. Rich.*) *Kokwaro* in K.B. 23: 530 (1969). Type: Ethiopia,
Mt. Scholoda, *Schimper* 51 & near Axum, *Schimper* 1489 (both P, syn. !, BM, BR, K,
isosyn. !) & several gatherings by Quartin Dillon (P, syn. !)

Umbellates (5–)7–16 per peduncle; mature spur plus pedicel (38–50 mm. long) longer than mature schizocarp (20–37 mm. long).

UGANDA. Karamoja District: Mt. Moroto, W. slope, 5 Sept. 1956, *Hardy & Bally* 10737! & Upe County, Loro, Sept. 1956, *J. Wilson* 251!; Elgon, Kapchorwa, 9 Sept. 1954, *Lind* 290!
KENYA. Northern Frontier Province: Mt. Nyiru, July 1960, *Kerfoot* 2039!; Naivasha District: Kedong Valley at foot of Rift, 22 May 1960, *Verdcourt* 2769!; Teita Hills, Ngangao Forest, 6 Feb. 1953, *Bally* 8758!
TANGANYIKA. Masai District: Kitumbeine, 11 Jan. 1936, *Greenway* 4310!; Pare District: Kiruru, May 1928, *Haarer* 1446! & near Vudea, 25 Jan. 1930, *Greenway* 2057!
DISTR. **U**1, 3; **K**1–4, 6, 7; **T**2–4; Sudan Republic, Ethiopia and Eritrea, also Mozambique and South Africa
HAB. Wooded and open grassland, bushland and thicket, extending to edges of and rarely into upland rain-forest, often in rocky places; 700–2800 m.

SYN. *P. multibracteatum* A. Rich., Tent. Fl. Abyss. 1: 119 (1847); F.T.A. 1: 293 (1868); Engl., Hochgebirgsfl. Trop. Afr.: 275 (1892); F.P.S. 1: 131 (1950); E.P.A.: 351 (1956); Fl. Pl. S. Afr. 20, t. 794 (1960)
P. usambarense Engl. in Abh. Preuss. Akad. Wiss. 1894: 61 (1894). Type: Tanganyika, W. Usambara Mts., Mlalo, *Holst* 363 (B, holo. †)

VARIATION. The petals over the range as a whole are generally white, but the variation from white to pink and dark red occurs quite commonly in East Africa. The dark red colour-form is most frequently encountered in the Kenya Highlands and in the Usambara Mts. of Tanganyika and has been described as *P. usambarense*. See also Müller in F.Z. 2: 147 (1963).

6. P. apetalum *P. Taylor* in Hook., Ic. Pl. 36, t. 3579 (1962); Müller in F.Z. 2: 145 (1963). Type: Tanganyika, Songea District, Matengo Hills, near R. Luhekea about 1·5 km. NE. of Mpapa, *Milne-Redhead & Taylor* 10377B (K, holo.!, BM, BR, EA, LISC, P, SRGH, iso.!)

Annual decumbent herb; primary stem ± ascending, branched; branches up to 1 m. long, usually slender (not more than 2·2 mm. in diameter), longitudinally grooved, sparsely pubescent and glandular, weak and straggling; internodes 2–6(–10) cm. long; vegetative parts, peduncle and pedicels glandular and pubescent; hairs patent, or somewhat appressed on the leaves; glands short-stalked to sessile. Leaves mostly opposite (one of every pair at each node usually smaller than the the other), those towards the root occasionally alternate; lamina deltoid to broadly ovate, 1·5–4·5(–6) cm. long and wide, subacute to obtuse at the apex, margin crenate, the base cordate, almost unlobed to shallowly 3- or 5(rarely 7)-lobed, membranous, hispid on both surfaces; petiole 0·8–4·8 cm. long; stipules lanceolate to ovate or deltoid, 2–4(–7) mm. long, 1–4 mm. wide, acute, often bifid, membranous, glabrous on both surfaces but ciliate along the margins. Inflorescence of 2 (rarely 3) flowers or the flowers frequently solitary; peduncle 5–10 mm. long, leaf-opposed or in the axil of the smaller of the 2 apparently opposite leaves; free part of the pedicel 3–12 mm. long; bracts 3–6 per pedicel, lanceolate or narrowly deltoid, 1·5–2 mm. long, 0·5–0·8 mm. wide, acute, membranous. Sepals narrowly oblong to lanceolate, 2·5–4 mm. long, 0·5–2·5 mm. wide, acute, pilose and glandular, usually pinkish in colour. Spur 1–3 mm. long, glandular. Petals absent or if present (as in Rhodesian specimens) not more than 2·5 mm. long. Stamens 9 or 10, all connate at the base for 0·5 mm.; antheriferous group 2–3(5), 1·8–2 mm. long; staminodes (5–)7–8, 0·9–1·2 mm. long; anthers as long as wide, 0·3–0·5 mm. Ovary usually 3–4-locular by abortion, with basal part tomentose; rostrum pubescent; style 0·2 mm. long, hispid to pubescent; stigmas 4 in Flora area, ± 0·5 mm. long, usually twisting around each other. Fruit 0·7–1·2 cm. long, hispid or pubescent; mericarps 2·8 mm. long, 1·2 mm. wide, sparsely pubescent. Seeds 1·8–2·4 mm. long, 0·8–1·2 mm. wide, glabrous; testa minutely reticulate, pale brown. Fig. 4.

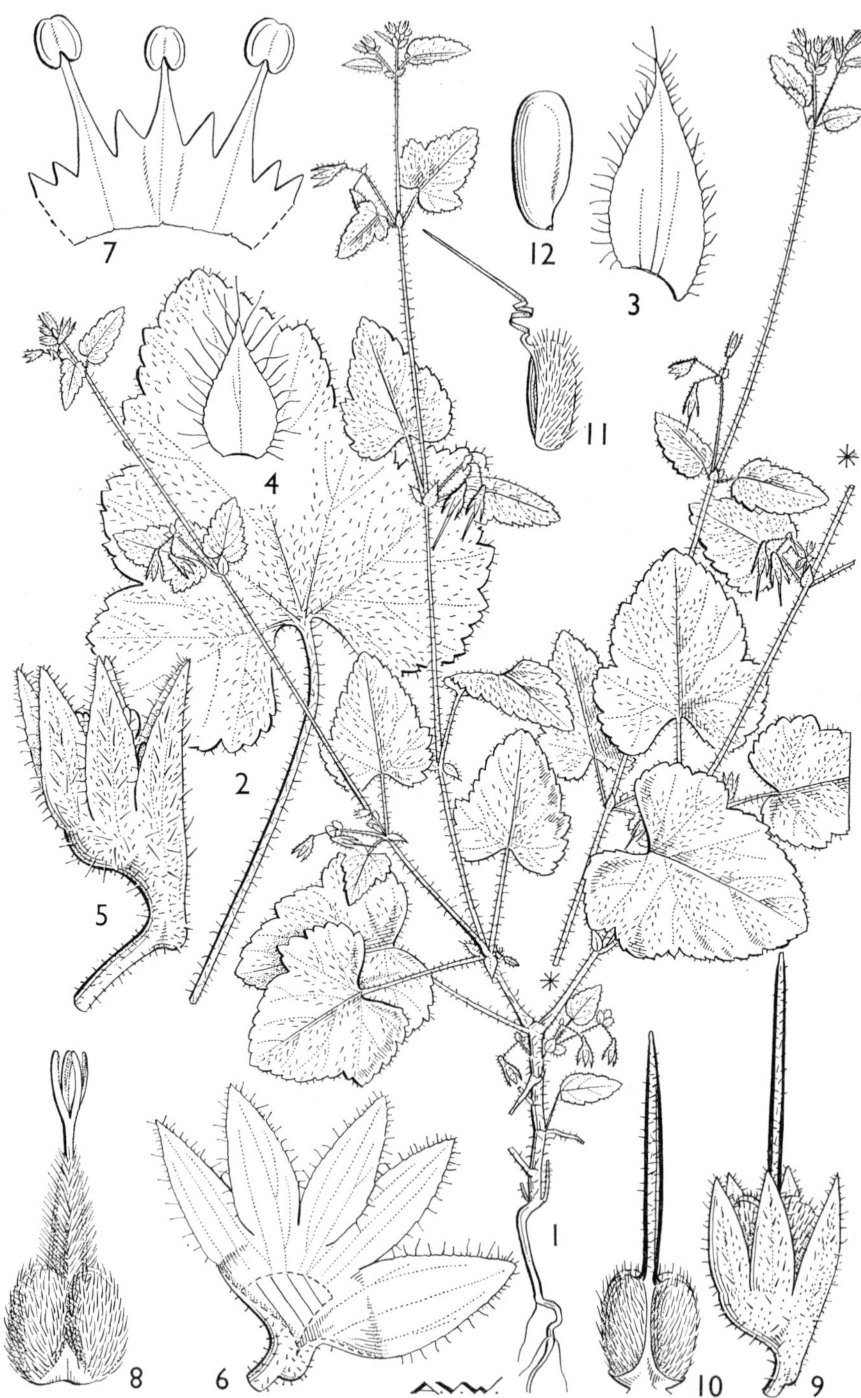

Fig. 4. *PELARGONIUM APETALUM*—1, habit, × 1; 2, basal leaf from a larger plant, × 1; 3, stipule, × 8; 4, bract, × 8; 5, flower, × 8; 6, calyx, opened out, × 8; 7, stamens and staminodes, opened out, × 12; 8, gynoecium, × 12; 9, fruit, × 6; 10, same, with calyx removed, × 6; 11, mericarp, × 6; 12, seed, × 8. All from *Milne-Redhead & Taylor* 10377 & 10377A. Reproduced by permission of the Bentham-Moxon Trustees.

TANGANYIKA. Songea District: Matengo Hills, Lupembe Hill, 20 May 1956, *Milne-Redhead & Taylor* 10377! & lower slopes of Luwiri Kitesa, 24 May 1956, *Milne-Redhead & Taylor* 10377A!
DISTR. **T8**; Malawi, Rhodesia
HAB. Upland secondary bushland derived from rain-forest, on hillslopes and rocky ground, becoming a weed in cultivation; 1500–2000 m.

INDEX TO GERANIACEAE

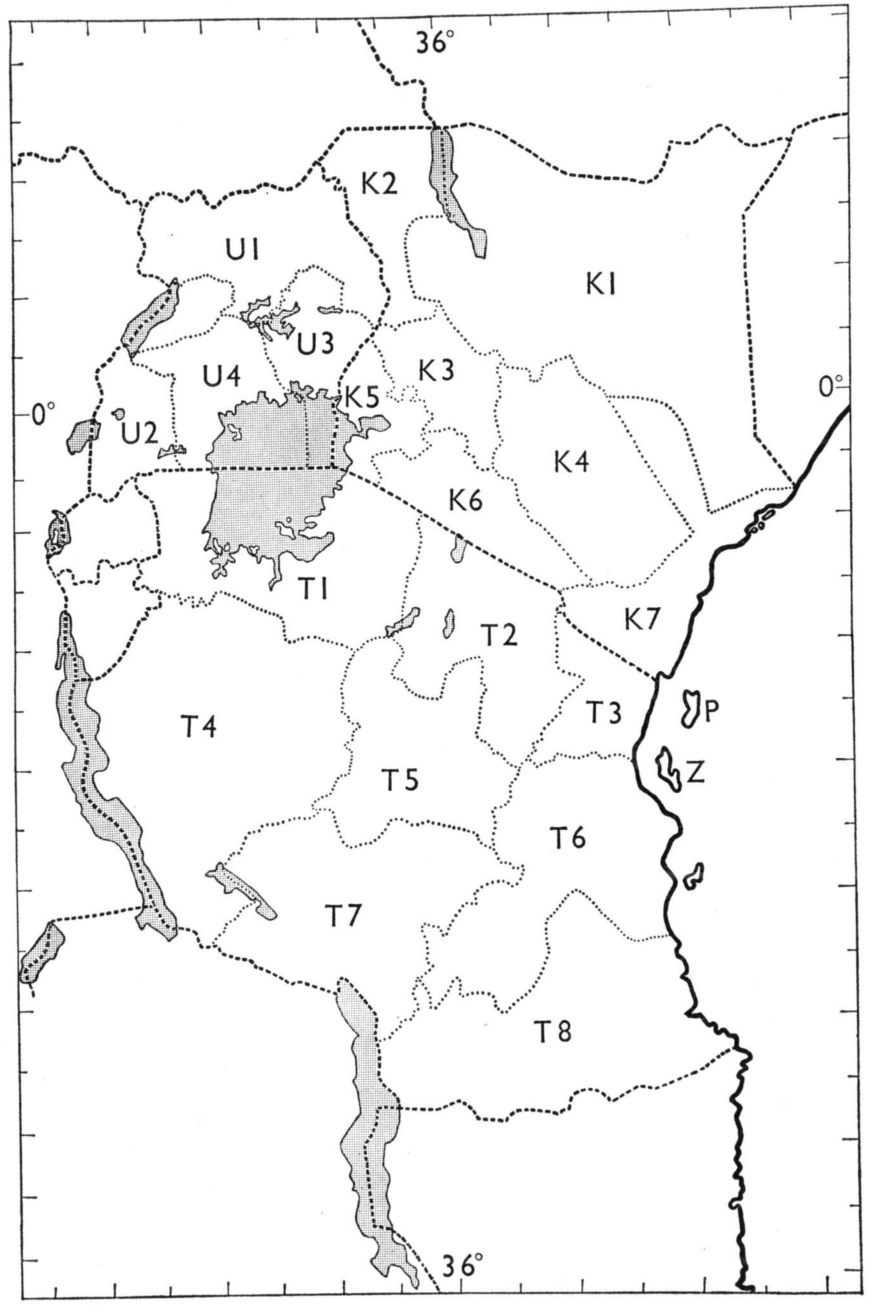

36°
K2
U1
K1
U3
U4
K3
K5
0°
0°
U2
K4
K6
T1
T2
K7
T4
T3
P
Z
T5
T6
T7
T8
36°